ABOUT HONEY

NATURE'S ELIXIR FOR HEALTH AND ENERGY

By

P. E. NORRIS

THORSONS PUBLISHERS LTD
91 St. Martin's Lane, London W.C.2

First Published	*February* 1954
Reprinted	*June* 1955
Reprinted	*September* 1956
Reprinted	*July* 1958
Reprinted	*April* 1959
Reprinted	*February* 1960
Reprinted	*January* 1961
Reprinted	*June* 1963

New Edition 1959

Made and Printed in Great Britain by The Lewes Press,
Wightman & Co. Ltd., Lewes, Sussex, England.

CONTENTS

ACKNOWLEDGMENT

I should like to thank Mr. C. C. Tonsley, Editor of the *British Bee Journal*, for placing his records at my disposal, and for the help he has given me; Mr. H. J. Wadey, of Beecraft Limited; Mr. G. A. W. Collins, and the Meadmakers Ltd., of Gulval, Cornwall.

CHAPTER ONE

WHAT IS HONEY?

IF YOU LOOK in the dictionary you will be informed that " Honey is a pale yellow, sweet, sickly, translucent, edible substance, provided by bees from nectar of flowers, and stored in a comb; and *virgin honey* is that which drips naturally from a comb." The word " honey " is derived from the Arabic " han," which is their word for the product. This became " honig," in German, and " hunig " in Old English.

The word has only pleasant associations. We describe something as being " as sickly as sugar," but " as sweet as honey," or " as honey to my soul." It has entered our language as a term of endearment.

Such people may exist, but I have never come across a person who dislikes honey, though most people have strong likes and dislikes about other edibles. Nor, so far as medical science has been able to ascertain, is anyone allergic to honey.

We know by analysis carried out in the laboratory almost the exact composition of honey. I say " almost," because although every constituent of honey can be accounted for by weight, there is some part of honey that cannot be weighed or measured, but has health-giving properties.

An average sample of honey, when analysed, consists of:

Water	17%
Levulose (d-fructose)	39·0%
Dextrose (d-glucose)	34·0%
Sucrose	1·0%

Dextrin	0·5%
Proteins	2·0%
Wax	1·0%
Plant acids (malic, formic, citric, etc.)	0·5%
Salts (calcium, iron, phosphates, magnesium, iodine)	1·0%
Undetermined residues (resins, gums, pigments, volatile oils, pollen grains)	4·0%
Total	100·0%

A word about sugars. This little book is not for the expert but for the normal, intelligent layman, who, when he reads books about food and health is baffled by the different names given to sugars.

Certain foods are known as carbohydrates. They provide the body with energy. There are three kinds of carbohydrates:

1. **Sugars**
2. **Starches**
3. Cellulose and related materials.

The sugars are Fructose, Sucrose, Lactose, Maltose, Raffinose, and Galactose. We are not concerned with maltose, which is formed naturally from starch during the germination of grain; raffinose, which is found in eucalyptus and cottonseed cake; or galactose, found in gums and seaweed, as they are absent in honey.

The sugars in honey are glucose, fructose and sucrose. Glucose is the simplest of the sugars. It occurs in the blood of *live* animals, in fruit and plant juices.

Fructose or grape sugar as it is sometimes called because it is also found in grapes, is a white crystaline substance which melts at a temperature of 187 degrees Fah. It is also called levulose and has the same number of carbon, hydrogen and oxygen atoms as glucose. But there are differences. Fructose crystallises more readily than glucose and is used by Nature for tissue building, whereas glucose in some way restores the oxygen that is replaced by lactic acid when fatigue sets in.

Chemically, sucrose is the same as cane or beet sugar; but it is a combination of glucose and fructose. When bees are fed mainly on white sugar far more than the normal amount of sucrose appears in their honey. Maple sugar is also rich in sucrose.

Dextrin is a gummy substance. Very little is found in honey, but it is this that makes honey so digestible, for dextrin goes almost directly into the bloodstream, and levulose does so only with slight modification.

The amount of protein found in honey is very small and when honey is filtered and re-filtered, to remove the cloudy appearance, the proteins, too, are drained away in the process.

Of the acids in honey, the mineral salts, and the undetermined residues, as well as the vitamins, more will be said later.

Now, it's a curious fact that though chemists have found out the exact composition of honey, they can't make artificial honey in the laboratory. Substances that appear identical to the naked eye have been made synthetically; but when fed to human beings they do not give the beneficial results that honey does.

There is some substance—no one knows what—that the bees have put into honey, that Man cannot. The secret eludes him.

One constituent is the vitamin. Until 1943 scientists said that there were no vitamins in honey, but experiments carried out by nutritionists proved these conclusions to be false. According to the *Swiss Bee Journal—1921* a number of children in delicate health were divided into three groups. One group was fed on the normal diet. The second was given medicines and body building foods as well. The third had honey added to the normal diet. After a few months this last group proved physically superior in every respect: blood count, weight, energy, and general appearance.

In 1943 Haydek, a scientist at the University of Minnesota proved that there are at least six vitamins in honey: thiamin (vitamin B1), ascorbic acid (vitamin C), riboflavin, panthothentic acid, pyridoxine, and niacin. Vitamin C is, of course, the anti-scorbutic vitamin, and the one that helps also to keep away the common cold.

But remember—honeys differ widely and some contain more vitamins than others. Also, the amounts of vitamin in honey are very small. Moreover, filtering honey through kieselguhr (earth) runs the vitamins away; and heating kills them.

As shown in the chemical analysis of honey, it contains numerous mineral salts.

In Austria before the war an experiment was carried out on 58 boys to find out whether the small amounts of copper, iron and manganese in honey would affect them. Each morning and evening 29 of the boys were given a teaspoon of honey, and at the end of twelve months a blood count was taken of both groups. Those fed on honey had gained a blood count of 8·5%, whereas the others were found to have lost 8·5%.

The mineral salts in honey form a small but important part of the whole.

Incidentally, if you prefer clear honey and wish at the same time to retain the vitamins, strain through a cheese cloth or fine metal strainer to remove coarse particles of wax or other impurities.

To make solid honey flow, heat it gently to not more than 104 degrees Fah., and never as high as 140. Avoid heating honey. Each time it is heated, some precious element vanishes.

And don't store honey in tins or metal containers, because the plants acids in it attack the metals. Bottle it in stone or earthenware jars.

Few people realise that, like wine, honey improves in flavour and aroma the longer it is kept. When T. M. Davies, the American archaeologist opened the tomb of Queen Tyi's parents in Egypt, he found a jar of honey, still in a partly liquid state. Though it had been sealed and placed there 3,300 years ago, the honey retained its full aroma and was as good to eat as when placed there.

Any bee keeper will tell you that good honey improves with keeping. An issue of *Nature,* 1944, contained a letter in which a Mr. Ronald Melville said that honey made in his Kensington apiary had a smell reminiscent of cats. This he traced to the pollen of the Tree of Heaven, *ailanthus altissima,* and the Sweet Chestnut, *castanca sativa,* so common in the London streets. The honey was tested and placed in jars in 1943, but by July the following year the unpleasant aroma had vanished and was replaced by a delicious, rich Muscatel. I tasted a similar sample of honey made in an apiary near Sloane Square, which, after being kept for two or three years, was rich and aromatic.

Honey is not a natural food in the sense that fruit and vegetables are. It is not, as some believe, the pure, undiluted secretion of flowers, gathered by bees and stored for Man's use.

Honey is a manufactured article, the bee being the manufacturing agent.

The odour and taste of honeys differ vastly, depending on the countries, and more often the areas in which they are gathered, and upon the soils in which the flowers from which the nectar comes, are grown. The smell, the healing power and the nourishing value of honey depends on many factors: rain, sun, mineral content of soil, and whether the plant is grown in humus or in artificial fertiliser. In this honeys differ every bit as much as coffees.

Before the war I used to buy 7 lb. tins of West Indian honey gathered in tobacco growing country. It smelt of fresh, fragrant tobacco and had a rich flavour all of its own.

When the nectar is gathered from the flower, three quarters of the bulk is pure water, in which about 20% cane sugar is dissolved, the remainder consisting of oils and gums that give it a distinctive flavour.

This is what happens. When the bee sucks in the nectar from the flower cup, the liquid passes into its honey sac and is mixed with acid secretions at the base of its tongue. The bee then flies back to the hive, but does not deposit her burden into the cells. She drops it, instead, into one of the honey houses, where it finds its way into the honey vats.

Some observers say that other processes take place, but as no one has ever looked into a hive while honey is being made, no one is sure what they are. A brewing process goes on inside the hive at a temperature of between 80-85 degrees Fah.—in the very heart of the hive it may be about 97 degrees—during which much of the water passes off in the form of vapour.

It is during this process that formic acid from the bee's glands is introduced into the honey as a preservative.

The bee is not a natural honey-maker. The young bees are taught their business by the more experienced ones, for they do not at first always visit the right flowers. I have watched bees by the hour—a pleasant if unprofitable occupation—and find that they will always fly to flowers of the same colour on any particular day, though this might not invariably be the case.

Bees will often fly two or three miles to find flowers with just the right kind of nectar.

Literally hundreds of species of flowers are visited by bees, but there is no honey in the world superior in taste and quality to that garnered on our English downs, and on the heather covered slopes of the Scottish highlands.

The amount of work that bees do in the harvesting of their honey is incredible. I have not worked it out, but those who have say that a honey-bee weighs one five hundredth part of a pound, and on each trip back to her hive she carries one half of her weight in nectar.

It needs ten thousand flights, therefore, for a bee to convey a pound of nectar to the hive—and nectar loses half its weight in evaporation.

As the average flight is about two miles, it is evident that a bee might travel as much as 40,000 miles to fill a single pound jar!

No wonder they call her the Busy Bee.

Much of our English honey is harvested from the clover, which yields about five pounds of pure honey per acre for each day the flowers are left standing in full bloom.

Well-known kinds of English honey are heather honey, which commands a high price, and for which there is a big demand; apple honey, and orange honey. Other popular kinds are brewed from sainfoin, bramble, rosebay and willow herb. Altogether there are more than 100,000 different flowers from which bees gather nectar. Nor could these plants breed but for the work of pollination carried out by bees.

Hives vary vastly in size from small colonies not much bigger than a football to the mammoth hive discovered some years ago by Dr. Gulmeth in the Australian bush. This was perched at the very top of a giant eucalyptus, more than 150 feet from the ground. The hive itself was 36 feet high, 21 feet across, and weighed a ton.

The eucalyptus was felled with great care so as not to injure the hive, from which Dr. Gulmeth salvaged 7,000 lb. of honey particularly prized for its medicinal value. When sold, his haul fetched more than £500.

Before the war Dr. Jaromir Rasin and his wife, noted Czech biologist, built a miniature concrete palace, surmounted by a cupola, in which lived more than 7,000,000 bees. Dr. Rasin's interest was not primarily honey, though he doubtless made excellent use of the bees' output. He kept bees mainly to study them and record his observations.

If you gain benefit from honey as a health food, you will as a natural course, grow interested in the way it is produced, the wonderful little cells that hold it, and ultimately in the bee itself. This may grow into a hobby and study that will continue for the remainder of your life, for bees have a curious fascination for those who watch them, and a small library has been written about the bee and bee society.

This booklet does not attempt to do more than whet your interest in the subject.

Everything the bee does is so mathematically correct and logical that I cannot close this chapter without saying something about the comb in which the honey is sealed.

The bee makes the wax for its comb, just as it makes honey. Wax is a by-product of metabolism. To make one pound of wax the bee has to consume six or seven pounds of honey.

Each cell is perfectly formed and consists of six sides, the angles of which are almost always the same, the only exceptions being cells fashioned to fill in odd corners. The hexagonal cell is made so accurately that the French scientist Reaumur proposed many years ago that Man could do worse than adopt the cell of the hive as a standard measure.

What is more, although bees have never studied Euclid, they will not use artificial hives if the cell measurements or the angles are the least bit out of true. Some years ago in New York a die-maker was commissioned to make a zinc die for pressing honeycomb into the foundations used in bee hives.

When the job was finished, the bees refused to enter the hive. The angles were checked and it was found that the die-maker, an expert in his craft, had been a fraction of an inch out—one part in many thousands.

How did the bees know?

But when new cells were made, they entered.

Bees also make their cells in such a way that they tip up ever so slightly, so that the honey will not drain out by gravity.

Nor will their cells melt until the temperature reaches 140 degrees Fah.; a temperature above the limit reached in any country in the world. The highest temperature ever recorded is that reached in Death Valley, Arizona—137 degrees.

CHAPTER TWO

SOME PERTINENT HISTORICAL FACTS

MOST OF THE classical writings contain references to honey. Shakespeare, who seems to have made succinct comments on every possible subject under the sun, mentions honey no fewer than 33 times; and honey-suckle, the bee and the honey comb on numerous occasions.

The ancient Greeks, Romans, Egyptians, Etruscans and dwellers of the Mediterranean referred to both the bee and their honey in their manuscripts; so did the Persians, Hindus and Chinese and all other races which range back to antiquity.

Being a food manufactured by the bee and there for the taking, it must have been one of the oldest edibles known to Man, who doubtless spied on the bear as he robbed the honeycomb, and followed his example—often with painful consequences.

All over the world there are rituals and ceremonies in connection with honey; so it would be strange, indeed, if the Bible contained no references to honey.

As early as 1707 B.C. Jacob sent Benjamin to Egypt with these words, " Take of the best fruits of the land in your vessels, and carry down the man a present, a little balm and a little honey, spices and myrrh, nuts and almonds ": Genesis 43.11. Honey was obviously considered among the best fruits of the land.

When about 200 years later David wrote a Psalm in praise of the Lord, he said, " The judgments of the Lord are true and righteous altogether.

More to be desired are they than gold, yea, than much fine gold: sweeter also then honey and the honey comb."

David waxed lyrical, setting a high value on honey. His son Solomon, however, was a sage and the very incarnation of wisdom. In that store of wise sayings, the Book of Proverbs (24.13) which has stood the test of centuries, Solomon advises, " Son, eat thou honey, because *it is* good, and the honeycomb, which is sweet to thy taste:

" So shall the knowledge of wisdom be unto thy soul."

Palestine was, and still is, a rich and fertile country. After their years in the wilderness the Israelites were so overcome by the abundance of everything it contained that they described it as " a land flowing with milk and honey," a phrase that is used a score of times in the Old Testament.

In the 14th Chapter of Judges we are told of Samson's encounter with a lion which " roared against him." And though he had nothing in his hand " he rent him as he would have rent a kid." And some days later when he came past the spot he saw—as was often the case—that bees had built a honeycomb in the carcase, so he took some of the honey, ate of it and gave some to his father and mother From that incident was born his famous riddle, " Out of the eater came forth meat, and out of the strong came forth sweetness." And after a bit of dirty work which enabled them to get the answer from his wife, his thirty companions gave the reply, " What is sweeter than honey? and what is stronger than a lion? "

There are authorities both modern and ancient who dispute this legend, for they say that bees dislike strong odours and decayed matter. But, in a country like Palestine, the bones of any dead animal left in the open would be picked clean by vultures in a matter of hours.

Honey, both wild and cultivated, was plentiful in Palestine. On one occasion after the Battle of Beth-aven, when the Israelites smote the Philistines hip and thigh and put them to flight, Saul ordered his people to eat nothing till evening. But, " all they of the land came into a wood, and there was honey on the ground," and Jonathan, who had not heard his father charging them not to eat, took some honey and ate it, and " his eyes were enlightened." (1 Sam. 14).

There has never been any doubt that honey has great sustaining power. When John the Baptist preached in the wilderness of Judea, we are told that " the same John had his raiment of camel's hair, and a leathern girdle about his loins; and his meat was locusts and wild honey." (Matt. 3.3 and Mark 1.6).

When the Children of Israel were making their long trek of forty years in the wilderness, manna was sent to them from heaven. It was a sustaining food, for manna alone kept them from starvation, and Exodus 16.31 gives us its composition. " And the House of Israel called the name thereof Manna: and it was like coriander seed, white; and the taste of it was like wafers made with honey."

In their wanderings the Israelites became a self-supporting race, turning all the scanty, hard-won materials in the wilderness to good use. And when they settled down and tilled their fields they developed an excellent system of husbandry; they had orchards, vineyards and colonies of bees. They not only sent their honey, which was famous for its quality, all over Asia Minor, but stored vast quantities, for they knew that honey improves with keeping.

That these stores were precious is recorded by Jeremiah 41.8, when Ishmael overcame and slew vast hordes of Israelites. " But ten were found among them that said unto Ishmael: 'Slay us not: for we have treasure in the field, of wheat, and of barley, and of oil, and of honey.' So he forbear, and slew them not among their brethren."

In a semi-tropical country, where food deteriorated rapidly, spices and honey, which are highly antiseptic, were valued highly. As valuable presents they took their place alongside jewels and gold, and we are told that when Jeroboam's queen went to see the blind prophet Ahijah at Shiloh, she took for him a cruse of honey (Kings 14.3). As honey is also used in parts of India and the Far East in cases of blindness, perhaps some of the queen's gift was placed on Ahijah's eyes.

As long as 3,000 years ago King David's Army had better rations than the British Army. They had " wheat and barley . . . beans and lentils . . . and honey and butter." (2 Sam. 17).

CHAPTER THREE

BRITAIN—THE ISLE OF HONEY

PERHAPS BECAUSE it is so much more costly than suger or treacle, honey is not eaten in as large quantities in Britain as it should be. Which is a pity, because in actual fact, comparing value for value, it is the cheapest food on the market.

When the ancient Phoenician traders came to Britain in search of tin and lead, they found such enormous quantities of honey eaten here that they called Britain the Isle of Honey. And the Druid bards called Britain the Honey Isle of Beli.

The face of Britain was then covered with forests with patches hacked out as common land, fields and orchards. Every village had its bees in skeps of wattled osier or woven basket work, and when the Romans landed in Britain they found that bee keeping was a national industry, and honey the universal sweetener.

When Plutarch landed in Britain he was astounded by the health and energy of the natives and wrote, "These Britons only begin to grow old at 120 years of age." One would hesitate to say that the secret was honey, but it is true that they consumed a great deal of it. And Pliny the Elder wrote, "These Islanders consume great quantities of honey brew."

Britain has always had a wonderful climate—though the weather may be foul—and in the days when the country was unpolluted by smog and undesecrated by hideous cities and streets where no grass or trees are to be seen, Britain must have been a veritable sanitorium.

In those days there was no sugar, which was a curiosity only for kings and the exceedingly wealthy.

Cane sugar was known in Europe about the First Century A.D. History tells us that Nearchus, Admiral of the Fleet of Alexander the Great, made an important voyage of discovery to India from which country he brought back a " honey-bearing reed."

Even the very rich did not use sugar till the Middle Ages, and it was not used by ordinary people till the middle of the seventeenth century. And then the sugar in general use was the very coarse variety now known in the trade as " foot-sugar," which is almost black and has a strong treacley smell.

Three hundred years ago honey was the only sweetener in the homes of poor and middle class Britons.

In the Middle Ages every Englishman from the King down to the lowliest commoner ate freely of honey, and to this must be credited in part the rude health and ability to endure hardship for which Britons were renowned.

The bee produced not only food and drink, but light as well, for the wax candle was the forerunner of the tallow candle.

Mead, a wine brewed from honey, was the traditional English drink, and from all accounts it flowed in rivers at every banquet. The main drink in the taverns was mead, though ale could also be had. In early times ale was unhopped, and before the reign of Edward VI, the use of hops in ale was illegal; but as late as the seventeenth century we find John Evelyn referring to hops as "that noxious weed."

Mead in those days, was King.

Incidentally, it was considered a crime for a priest to enter a tavern, but this, apparently, was little hardship, for every monk had his fair ration, and we are told that Ethelworld's monks was each allowed a sextarium of mead for dinner, a measure amounting to several gallons.

In the course of time bee-masters became renowned for their wonderful brews, many of which had health-giving properties, and during the sixteenth and seventeenth centuries mead brewed according to secret recipes was indistinguishable from the very finest Canary sack.

It had a pale golden colour, was full of air bells and bubbled like champagne.

Meads are not peculiar to Britain; they have been brewed by primitive tribes all over the world since the dawn of time. Some are highly alcoholic; others are no stronger than pure apple juice. But all are supposed to be imbued with health-giving properties. And it is one of the curiosities of philology that all over the world the name has a similar ring. In India it is *madhu*, made from *modhu* or honey; in Scythia, *medos*; in Greece, *methu*; in Old Irish, *mid*; in German, *meth*; in the Slav countries, *medu*; in Lithuania, *medus*.

In Britain there were many varieties of mead renowned for their invigorating qualities. Morat was made with honey, water and the juice of mulberries; piment or clarre, Chaucer's wine, was a mixture of honey, spices and wine; bracket, another popular drink, was made of honey, wort and spices. Piment was served at important banquets and was popular with kings and nobles.

The most renowned health-giving of the meads was hydromel, brewed by a secret formula and dispensed in Ancient Greece by the great Galen. Pliny records that it could modify anger, sadness and affections of the mind, which makes one suspect that it was slightly alcoholic. He writes, "It is an extremely wholesome beverage for invalids who take nothing but a light diet; it invigorates the body, is soothing to the mouth and stomach, and by its refreshing properties allays feverish heat. It is well suited to persons of chilly temperament or a weak and pusillanimous constitution, diminishing also the asperities of the mind."

Other renowned honey beverages were ompacomel, made from fermented grape juice and honey, and taken as a soporific; oenomel, pure grape juice and honey, used for nervous disorders and gout; conditum, honey mixed with wine and pepper, supposed to alleviate digestive upsets; oxymel, honey, vinegar, sea-salt and rainwater, an infallible specific for sciatica, gout and rheumatism (also used as a gargle); and clysma, a concoction of honey and water, which cleansed the bowels.

The muslum of Ancient Rome, and the lipez of Russia, are also famous honey wines.

Honey vinegar is used, even today, as a salad dressing, and is made from inferior grades of honey. Five parts of water to one of honey are used. The mixture is then boiled in an earthenware (never metal) dish, and a little yeast added. Place the barrel—or bottles—in a warm room and cover the bung with a cheese cloth. The vinegar should be ready in a few weeks.

Apparently there is a good deal of virtue in mead, for honey contains malic acid which counteracts gout and rheumatism, and this is transferred to the brew. Cider, incidentally, also contains malic acid, and that is why it is often recommended as a specific to rheumatic sufferers.

Beck and Smedley, in their excellent book, *Honey and Your Health,* write, "For many centuries mead was considered a veritable *elixir vitae.* Its principal medicinal value was in kidney ailments, as an excellent diuretic without disastrous effects upon the kidneys. As for gout and rheumatism, mead ranked not only as a curative but also as a preventive medicine. It was widely used as a good digestive and laxative."

One of, if not the greatest living authority on mead-making is Mr. G. A. W. Collins, of Ruislip, Middlesex. For years he has combed the libraries and ancient manuscripts for recipes, and today he is considered the High Priest of the cult of mead brewing.

Mr. Collins has not commercialised his secrets, but he sends his recipes to those who wish to know how to make mead at home.

The only man I know who has put mead on the market is Lt.-Col. G. R. Gayre, M.A., D.Sc., of Gulval, 12 miles from Land's End. He claims that his recipes are centuries old, too, and brews from honey a sherry-type aperitif known as *cyser*; a herbal liquor called *sack methaglin*; a dry table wine called *mead*; and a dessert wine known as *dessert mead.*

Both Mr. Collins and Lt.-Col. Gayre claim that mead is superior to grape wine, for it leaves no hangover, is a blood-cleanser and acts as an invigorating tonic.

The Gulval mead-brewery uses 250,000 lb. of honey every year and produces the equivalent of a million pounds of grape juice.

Incidentally, the authorities in the U.S.S.R. recently stated that of 200 people who were authentically found to be between 110 and 150 years old, nearly all were or had been beekeepers and ate a fair amount of their own produce.

CHAPTER FOUR

HONEY FOR ENERGY

HONEY IS ONE of Nature's finest energy-giving foods.

The danger of writing a sentence like that is that those who know nothing about food—and their name is legion—begin to fill themselves with honey till they are heartily sick of the taste of the stuff; or take so much of it that they harm themselves. For one can take too much of any good food—even honey.

Having written this, let me tell you why it is good for energy and why it combats fatigue. But, first let me ask, " What makes a muscle tired? "

When a muscle is exercised in vigorous sport or hard labour, it rebels. This is because a substance called glycogen in the muscle disappears and is replaced by an equal quantity of lactic acid. Pre-formed carbon dioxide is driven off and heat in proportion to the amount of lactic acid is formed. Also, the hydrogen ion concentration rises, contracting the muscle.

And for every gram of lactic acid formed in the muscles, the body gives off 370 calories, or units of heat.

Glycogen is a substance that is quickly transferred into heat and energy, and as fatigue sets in, this store of glycogen in the muscles is used up.

By experiment it has been found that if a muscle is left in hydrogen, it stiffens and contracts; but if left in oxygen, it remain soft and supple.

During exercise or labour an oxygen debt is built up and one of the ways to pay off this debt is to rest, during which period generous quantities of air are taken into the lungs, the oxygen debt paid off and the lactic acid dissipated.

But rest is not always possible.

The heat—370 calories to every gram of lactic acid—has vanished, too, and must be replaced.

Carbohydrate is the primary fuel of muscle, and the most easily digested form of carbohydrate, and one which because of the dextrin it contains can be assimilated into the bloodstream directly and without going through the complicated processes of digestion, is honey.

A pound of honey, when eaten and converted into carbon dioxide and water, generates about 1,450 calories.

Now we know through experiment and research that the heart uses almost as much oxygen as the voluntary muscles, but unlike them is unable to stop beating and have a good, long rest. If your arms and legs ache from fatigue and you keep going through an effort of sheer will, there comes a time when your body will give up—must give up—and rest.

But, whether you drop from fatigue or not, your heart keeps on beating. If it doesn't—you've had it.

One of the main factors of stamina, therefore, is a strong heart. *And doctors say that honey is one of the finest of all foods for the heart!*

They are constantly finding new uses for honey in their fight against disease, and there does not seem to be an organ in the body that does not respond favourably when honey is eaten.

Honey is strongly recommended by Lelord Kordel in his book *Eat and Grow Younger,* and by that eminent nutrition expert, Dr. Arnold Lorand, who in his work *Old Age Deferred,* says, " It has been found that the heart of animals removed from the body will survive for days if kept in a salt solution with grape or fruit sugar added. The addition of certain mineral salts like lime and carbonate of sodium is also able to prolong the survival of the cut-out hearts of dead animals. So there can be no doubt that the same elements must also prove useful to the heart of the living, as is indeed the case.

" The ingestion of sweets promotes muscular activity and fatigues from bodily exertion are better borne. And, this also holds good for our most important muscle, the heart.

" I have seen in my heart patients very good results through the addition of a generous amount of sweets to their ordinary diet. On the other hand, I have, as a rule observed a weak activity of the heart with my patients in Carlsbad suffering from graver forms of diabetes who were kept on a diet strictly excluding sweets and starchy foods in general.

" Indeed, a weak heart is most frequent in severe diabetes, as in such a condition the sugar ingested cannot be utilised and is entirely eliminated in the urine.

" For this reason I consider it unwise to place severe cases of diabetes on a strict diet and I recommend them to use the fruit sugar (levulose), which is often well utilised,, and especially in a case of diabetes with heart failure, I like to do this. Such persons should never be strongly dieted.

" As the best food for the heart I recommend honey on the base of the above mentioned observations. Honey is easily digested and assimilated; it is the best sweet food, as it does not cause flatulence and can prevent it, to a certain extent promoting the activity of the bowels. It can easily be added to the five meals a day I recommend in cases of arterio-sclerosis and of weak hearts.

" As it would be unwise to leave such a hard working organ as the heart without food over the long hours of the night, I recommend heart patients to take before going to bed a glass of water with honey and lemon juice in it, and also to take it when awakening at night.

" Before and after muscular exertion honey should be given in generous doses; no coachman would allow his horses to run for hours without giving them food at the resting intervals. Only man is so unreasonable as to undertake heavy exertions with an empty stomach. No wonder that so many sportsmen get a weak heart simply for just such a reason.

" The use of cane sugar cannot replace honey. In the same amount it is chemically irritating to the stomach."

Everyone who has read books dealing with food, dietetics and the metabolism of the body has heard the phrase " blood sugar level." This simply means that a certain amount of glycogen must be present in your blood all the time. Otherwise your efficiency decreases.

Your brain may slow down and you may fall into a coma.

In a less extreme case you would merely feel a desperate lassitude. And when you feel like that your mind turns instantly to those well advertised synthetic glucose foods that are supposed to restore dissipated energy.

What happens is simply this. The sugar content of your blood falls so low that extreme fatigue sets in, and your heart, a powerful muscle, slows down.

If, in such a condition you eat a meal, there is an almost instant reflex action and you feel slightly better, but your body does not get any real energy. It has to wait till digestion takes place and some of the food is converted into dextrose, which in turn produces the glycogen that gives you energy.

So, if you wish for an instantaneous shot of energy; one that can be felt in about ten minutes, eat a spoonful of honey. The glycogen passes rapidly into your bloodstream and has the same effect as a meal.

According to Dr. Christopher Woodward, Honorary Consultant to British Olympic Teams—1949 and 1952—

author of "Sports Injuries," "In some games, particularly squash rackets, one exerts oneself so furiously that one becomes short of sugar, developing a feeling known as 'the shakes.' This is due to the fact that the sugar in the blood has dropped below a certain level. I have said elsewhere that the body cannot digest food and take vigorous exercise at the same time, but sugar is something that you can take in such a form that it can be easily absorbed and its effect felt almost instantaneously. The best forms of sugar are brown sugar, honey, treacle and white sugar. I do not recommend glucose because it usually gives fit men and women indigestion. It is, so to speak, a pre-digested form of sugar, and whereas it is all right for old ladies who cannot digest sugar for themselves, it is not the sort of thing that a normally fit person should need to take.

"You may find, however fit you are, that you do not digest even sugar very easily; in that case, if you develop the shakes you may well have to stop playing for a few moments while the sugar takes effect. If in training you find that you have a habit of developing the shakes, you should take extra sugar an hour or two before playing; this will enable you to get through any match without being affected in any way. 'Know thyself,' is the way the Greeks put it."*

There is no energy producing food on the market to touch honey. Nor will there ever be, for Nature knows more than the cleverest chemist. All synthetic glucose products are inferior to honey, not only for the speed with which they bring about a sense of well-being, but also for the production of real lasting energy and the alkalinising of the body. They are a sheer waste of money when you can buy Nature's genuine energiser.

This does not mean that you can exist on a diet of honey

* *Quoted by courtesy of the publishers, Messrs. Max Parrish & Co. Ltd.*

alone, or fly to the honey pot the moment you feel washed out.

Remember this. It is important. You can't eat the wrong foods most of the time, pour injurious liquids down your throat, burn the candle at both ends and expect an occasional spoon of honey to make up for bad living.

Learn the art of healthy living and eating. Take adequate doses of rest and refresh your body and mind by recreation and play. And then, if you overdo it occasionally, honey will replenish your lack of energy.

But don't attempt to rely upon it as you would a drug. It is not a panacea for all ills.

There is no one wonder-food or drink in the world. Anyone who tells you so is a charlatan and a scoundrel.

Energy is generated mainly by the carbohydrate foods, of which honey is one of the most important because of its pleasant taste, the wide variety of uses it can be put to, and the fact that it is so easily assimilated.

But in order to get the utmost benefit from your food, your meals should be balanced. That is, you should eat sufficient proteins to repair cell wastage; fats for warmth and energy; and fruits and green foods—cooked and uncooked—for vitamins and mineral salts, and for certain substances that the scientists have not yet isolated.

There are certain substances in fresh foods that have not yet been isolated. What they are, no one knows. After all, it was not till 1911 that Sir Frederick Gowland Hopkins pinned down the first vitamin, and ever since, scientists have been finding new ones.

There is certainly something vital in honey that has eluded the analytical chemist and the research worker, something they have not been able to put their fingers on, for though they can reproduce faithfully every constituent of honey, the bee still has the final laugh.

One of the great champions of honey was the late Dr. Josiah Oldfield, who lived as active and varied a life as any man.

This remarkable man went to Oxford on a scholarship which just about paid his fees and allowed him to live. He had no money for the things that make life worth living, so while he was a student he got a job with the local council, breaking stones on the roads! Every morning he was up at five and on the job by six and had completed half a day's work by the time other students started on their studies. " And what is more," he exults, " I was paid for taking this form of healthy exercise! "

And the food that kept him going and enabled him to do two jobs without breaking down, was vegetarian: fresh greens, fresh fruit, nuts, milk and honey.

" There is no food," wrote Oldfield, " that surpasses honey for all ages of the human race.

" Especially when people have passed middle age the organs of the body are able to replace the energy obtained in youth from many and varied sources by getting one of the highest and most valuable foods, *i.e.* that obtained from what the Bible calls " honey and the honeycomb."

" I would like to impress upon all readers who lack vitality or who have digestive difficulties that they will find the daily addition of good honey to the dietary both a pleasure and increased physical strength, and will do away with many digestive troubles if they will replace the flesh part of their food by a daily tablespoon of English honey."

Oldfield knew what he was talking about, for though he went to Oxford to become a Doctor of Civil Law, he took medical degrees—L.R.C.P. and M.R.C.S.—at London.

He is not the only medical man to recommend honey for failing energy. Dr. Reginald Austin, who at 70 carried out a 30-day fast to rejuvenate his body and had the energy of a young man, was a great believer in the use of honey. " In place of jam," he used to say, " eat honey; honey with milk; honey with ground nuts; honey in salad dressing. Use it in place of sugar."

Athletes all over the world have found that honey revives flagging energy faster than any known food, and unlike other synthetic glucose foods, has no ill effects after. When in July 1951 Philip H. Rising, of Rotherham, swam Lake Windermere—10½ miles in 8 hours—he was fed throughout on honey and treacle sandwiches.

One of the greatest feats of endurance in modern times was the scaling of Everest by Sir Edmund Hilary and " Tiger " Tensing. Sir Edmund is a New Zealand bee-farmer, and after his climb his father is reported to have said, " It was honey that did it." Tensing, too, is a honey-eater as are most Sherpa porters who eat a good deal of honey, which is abundant in the Himalayas.

The citizens of ancient Athens were as fit and strong as people ever have been. They ate little and their diet consisted almost entirely of cereals, figs, a little fresh fruit, salt fish, cheese, milk, yogurt, honey and olives. Their athletes supplemented the diet with honey.

Jim Londos, world heavyweight wrestler in the nineteen thirties, was a Greek. He was about five feet eight inches tall, and beautifully proportioned. His diet was the old Greek one—with plenty of honey.

The greatest of all Indian wrestlers, Gama the Tiger, the man who put the giant Pole, Vladivas Zbyscko on his back in a matter of seconds, ate a great deal of honey in milk and sweets every day. The sweets favoured by all Indian wrestlers and strong men—*labri, ludoos* and *saundaesh* should all be made with honey or molasses and cream, or honey or molasses and yogurt.

One of the most enthusiastic advocates of honey is Bernarr Mcfadden, Father of Modern Physical Culture. For more than half a century he has written and lectured about food and advised the use of honey in place of other sweeteners.

He is, of course, a crank, as are all with original ideas and minds of their own. Without a crank, no machine could move. The human crank is the man who gets things moving.

Macfadden carried out hundreds of experiments with his own body, to find out what it could and could not do; to discover how little he could live on without falling ill; to discover what was the finest food for giving energy. At one time he walked bare-foot every morning from his home in the country to his office in the city, a distance of 23 miles. And all the sustenance he needed was a glass of water to which a spoon of honey was added.

This remarkable man married again when well over 80, and makes a parachute jump on every birthday, just to show that there's still a kick left in him.

Incidentally, honey is one of the cheapest foods, weight for weight, known. Seven ounces of honey are, according to the *Bulletin of the Food and Dairy Dept.*, Iowa, the equivalent of: 1 quart of milk; 5-6 oz. cheese; 10 eggs; 12 oz. round beefsteak; 15 oz. boneless cod; 8 oranges; 8½ oz. walnuts.

A United States Ministry of Health Report—No. 9 of 1936 gives the energy value of a pound of local honey as being equal to 1,290 calories, and a pound each of eggs, bread and lean meat as 1,200, 1,280 and 1,150 respectively. The Ministry claimed that one pound of honey is equal to 30 eggs, 6 pints of milk, 8 lb. of plums, 10 lb. green peas, 12 lb. apples or 20 lb. carrots.

One must accept all such tables with reserve for the calorific value not only of honey, but other foods varies with the kind of soil it is grown in, the season of the year and other factors.

Because of its high calorific value and the ease with which honey passes into the blood, it is one of the best foods for hikers, campers and above all, mountain climbers. In *Lords of the Rockies,* Wendell and Lucy Chapman say, " We do not use any stimulant except weak tea with a *tablespoon of honey per cup.* After a 20-30 mile hike a pint of this pick-me-up revives us completely in 15-20 minutes with no let down afterwards."

The great W. G. George, holder of the world's mile record for 27 years, was not a food faddist and often drank a bottle of beer and smoked a cigar half an hour before a race; but he was a great believer in honey and ate a great deal of it. He liked it for its pleasant taste and believed that it gave him just that extra energy for all his great efforts.

CHAPTER FIVE

HONEY AS A HEALER

I SHOULD NOT LIKE you to run away with the idea that honey is either the finest healing agent in the world, or the only one. But it has such great soothing and healing power about which so little is known, that it would be as well for you to know what it can do. For, whereas most of us do not stock medicine cupboards, there is usually a pot of honey in the larder.

The ancients valued it not only as a sweetening agent. They lauded its many virtues. The Hindus used it as a mild laxative and wild honey kept for 12 months was, and is still used as an astringent.

Swami Sivananda claims that " it will strengthen a weak heart, a weak brain, and a weak stomach." Most of us suffer from one such weakness, if not all three.

He continues: " Honey is a heart stimulant. It is useful in cases of malnutrition. It should be given for general physical repair. Honey kills bacteria and thus enables the body to overcome diseases. Disease germs cannot grow in honey. Honey is substituted for orange juice and cod liver oil. It is useful in bronchial catarrh, sore throat, coughs and colds. Diabetic patients can also take honey with advantage.

" Honey can be taken with milk, cream or butter. It is a restorative after serious illness. It enters into the combination of Chavan Prash and other *kalpas* (Ancient Hindu medicine). As soon as a child is born its tongue is smeared with honey. This is the first food that the child takes.

"Honey is more stimulating than alcohol. If you take a tablespoonful of honey in hot water when you are tired or exhausted by over-exertion, it will brace you up immediately. Soak ten almonds in water overnight. Remove the skins in the morning. Take them with two tablespoons of honey. This is a brain tonic."

Swami Sivananda is a Yogi, and in saying this he draws on centuries of Yoga knowledge, for honey is one of the foods mentioned in the Yogi books, as having the power to cleanse and revitalise.

Every ancient religious book from the *Rig-Veda,* incribed in Sanskrit roughly 3,000 years ago, to the *Book of Mormon,* 1830, praises the bee and its product, honey. That they all do so is no coincidence.

In Ancient India the bee was never farmed, as in Europe. Hives were unearthed in the jungle, the bees smoked out and most of the honey drained off. If men failed to rob the bees, bears climbed the trees and devoured it, for curiously enough, Nature has given jungle bears such thick, close fur that bees cannot penetrate it and sting them.

For centuries a mixture known as *ceromel,* consisting of one part of bees' wax to four of honey, has been used in the treatment of ulcers in India by the *kavirajs* (healers).

Abyssinia is another country where wild honey has been widely used for centuries. A very potent mead is made from honey and carried by travellers in huge horns. Hunters sometimes chase their prey for days, subsisting exclusively on this beverage, which energises but does not intoxicate.

In ancient Russia they had a drink called lipez, made from honey brewed by bees from the nectar of the linden flower –and very highly prized it was.

One of the noblest Roman drinks was muslum, made from honey, wine and water boiled, which was praised by gourmets and connoisseurs of food.

Honey has always been a treasured food, even before Glaucus, son of Minos, was smothered in a cask of honey.

Bees, too, have been praised for their industry. The ancient Egyptians symbolised their kings under this emblem. Honey was the reward given to the meritorious, and the sting was the punishment of the unworthy. The bee was also the emblem of the ancient French empire, and the royal mantle and standard were embroidered with scores of raised bees in golden thread instead of "Louis flowers." More than 300 golden bees were discovered in the tomb of Childeric in 1653.

The bee symbolises industry and his output the most precious of all things; hence Sophocles was known as the "Bee of Attica," and Plato the "Athenian Bee."

The bee was so revered by the ancients that the Koran, the holy book of the Moslems, devotes an entire chapter to the bee, which according to Mohammed was the only living creature ever spoken to by God.

According to an Arabic scribe the Prophet is supposed to have said, "Honey is a medicine for the soul; benefit yourselves by the use of the Koran and of honey." And even today, when devout Moslems all over the world eat honey, they murmer before the first drops touch their lips, "Bism Allah" (in the name of Allah), or "Allah Akbar" (Allah is Great).

When the first Mormons went to Utah they took with them the deseret or honey bee, and they intended to call their state Deseret instead of Utah when they were first admitted into the Union. Today Utah is known as the Beehive State and has for its seal a beehive surrounded by flowers, surmounted by the single word "Industry."

All references to bees in ancient writings are laudatory. The Indian love-god Kama's bow string was formed by a chain of bees and according to Greek mythology, Zeus' sole food was honey.

The Incas, Aztecs and Zapotic Indians knew of some of the many virtues of honey and used it freely, not only as food, but as antiseptic, and mixed it into their anaesthetics when they carried out some of their remarkable dentistry. They gave their patients a mixture of wild mushrooms and honey before tackling the more difficult operations. It is difficult to see how some of their fine gold inlays could have been done otherwise.

In Scotland centuries ago, a wasting disease was cured by a concoction known as Athole Brose, which consisted of equal parts of good thick honey, preferably from the ling heather, and mature Scotch whisky from the pot still. Little and often was the old wife's advice—and the cure had naught to do with faith.

And if folklore has anything to it, honey was also a sure cure for baldness. If massaged well into the scalp it caused the shiniest pate to sprout hair; but no assurance can be given in this work that honey will cure baldness.

Honey, because of its powers as an antiseptic and a preservative, was used in Ancient Egypt, China and all over the East for embalming. It still is used in Burma for the same purpose. No European who knows that country well will readily buy honey unless he is sure of its source—for who knows—it might have been used to preserve a body! The *pongyis* (priests) often refuse to bury a man till the full costs of the funeral have been paid. If the money is not forthcoming, they place the body in honey. When they get their fees the honey is drained or scraped off and placed in earthenware jars, and if no longer wanted, is taken to the market place and sold.

But so powerful is its antiseptic quality that even this honey is germ-free and hundreds of Europeans who have eaten it without being aware of the use that it was put to, have suffered no ill effects.

But the very idea of putting the stuff on one's bread might not appeal to the hyper-sensitive.

All over the East honey has been used, and is still used as a preservative for cakes, sweetmeats and other delicacies in the same way as spices are used to form the base of curry powders, to preserve meat and vegetables. Because of the intense heat, food decomposes rapidly and in the case of meat would have to be thrown away in a few hours; not much longer in the case of cooked vegetables. From shear necessity the peoples of the East have had to find preservatives; and honey and curry powders are the result.

In India honey and milk, or honey and tyre—yogurt—is given as an offering to honoured guests and to bridegrooms; and throughout India and the Arab countries a mixture of cream, fresh butter and honey is a favourite dish.

It is no accident that honey has been used as a cleanser and a purifier in religious rites in countries thousands of miles apart.

The lore of the ancients has been borne out by modern physicians. Before the war Dr. N. Zaiss, a leading Viennese physician subscribed a paper to a medical journal in which he claimed to have treated thousands of cases of wounds, sores and long-standing ulcers with honey, and in not one case did he fail. Honey, he said, soothed pain, acted as an antiseptic, hastened healing and was specially effective in curing burns and carbuncles.

CHAPTER SIX

HONEY—THE MEDICINE

CLAIMS, in many cases substantiated, have been made for honey as a medicine. There are people who greet all such claims with derision and ask, " Then, why doesn't the medical profession use it? "

The answer is that until recently most doctors did not know of the therapeutic value of honey; nor for that matter did most of them know that salads and fresh fruit and rich brown bread were better than the starchy, soggy messes and white bread on which they fed their patients for so long. Even today hospitals give their patients white bread; but that does not make white bread superior to wholemeal.

Until comparatively recently, for instance, very few people except a few old countrymen to whom the secret had been handed down through generations, realised that honey gave relief to sufferers of hay fever and asthma.

Just before the war a German doctor experimented and found that if you know the kind of pollen that affects your hay fever, and can get honey made from the nectar of flowers that contain that pollen, it will relieve and eventually cure your hay fever. Doctors innoculate you with a serum made from that pollen; Nature does the same thing in a much more pleasant way.

The ancients said that Nature's remedy for every evil lies close at hand, as in the case of the nettle and the dock leaf, but Man is too obtuse to look for it.

Doctors tell us that there is more than a passing connection between hay fever and asthma; both complaints seem to affect highly strung people. But we know, because it has been tried often enough, that if a jug of honey is held under the nose of an asthma patient and he inhales the air that comes into contact with the honey, his breathing becomes deeper and easier.

If you suffer from short-breath or asthma, try this. Many asthma sufferers keep jars of honey at hand, and though the effects last for only about an hour or so, relief is often instantaneous.

Why is this so?

German scientists investigated the problem and came to the conclusion that honey contains a mixture of " higher " alcohols as well as ethereal oils, and the vapours given off by them—often imperceptible to the healthy—are soothing and beneficial to the asthmatic.

The best honey for the asthmatic is that harvested by bees on the slopes of hills on which conifers abound. The balsams and turpentines give honey its healing tang.

There are other constituents; secretions of the bees, perhaps; hormones which have hitherto defined analysis.

Honey is also invaluable in cases of all pulmonary diseases; that is, diseases of the lungs. Whether the air flowing over the honey is inhaled; or whether it is eaten plain or taken either in milk or water, it usually brings relief.

Honey is found in cough mixtures and there is no known cough mixture or linctus to equal a few drops of ipeccacuanha wine, a teaspoonful of lemon juice, and a spoon of honey.

An old Indian recipe for asthma, for instance, urges you to take an egg, place it in a wine glass and cover it with lemon juice. Let it remain there for 12 hours, by which time the shell will soften. Then remove the shell and beat yolk and white together. Add a quantity of honey equal in volume to the lime or lemon juice and mix. Bring to the boil and take off the fire when it starts to get thick. Cool off and bottle. Take a teaspoonful, night and morning.

Most sweeteners—particularly white sugar—irritate. Honey soothes. In 1939 the Red Cross Hospital at Hamburg produced an ointment from liver and honey which was applied with immense success to burns and furuncles. The doctors declared that the healing power of the ointment was due almost entirely to the antiseptic properties of honey.

The *Zeitung fur Aerztliche Fortbildung* of that period informs us that experiments carried out with 15 kinds of honey showed that they killed bacteria of many kinds in small quantities. The report stated that other honeys might do likewise, but only 15 varieties were tested.

The scientists said that the antiseptic property of honey came from the various acids in it: formic, malic and lactic, though this does not explain all.

They also said that fresh honey was more effective than that which had been kept for a long period.

During the war the Admiralty appealed to bee keepers asking them to send their surplus honey to submarine crews because (1) the rich taste of honey would vary the monotony of their rations, (2) it was a good food in concentrated form, (3) it is a substitute for smoking, which, of course, is not allowed in submarines. Thousands of pounds of honey were sent to *Sea Lion, Sea Wolf, Lion, Graph, Sturgeon, June, Unbroken, Tribune,* and *Upstart.*

Sir Winston Churchill, as a countryman, realises the value of bees as pollinators and honey as a food. In 1943 Mr. F. Austin Hyde, Headmaster of Pickering Grammar School, a renowned apiarist and an authority of North Riding dialect, wrote to Mr. Churchill asking for a bigger ration of sugar for bee keepers. The letter fell into the hands of the Hon. William Mabane, Private Secretary to the Ministry of Food, one of Mr. Hyde's former pupils, and he brought it to the notice of the Prime Minister, who despite the great burden on his shoulders thought the matter important enough to write to the Food Minister, saying, "What is the amount of sugar issued to bee keepers, and what is the saving in starving the bees of private owners?"

As a result, bee keepers got a bigger issue of sugar, the bees were enabled to go about their business of helping to produce the nation's fruit crop, and when Mr. Hyde came up to London for the National Honey Show he left a pot of the finest Yorkshire honey at 10 Downing Street, "with the gratitude of the bees."

I have mentioned that the ancients used honey as an antiseptic. St. Ambrose said, "The fruit of the bees is desired of all, and is equally sweet to kings and beggars, and it is not only pleasing but profitable and healthful; it sweetens their mouths, cures their wounds, and convaies remedies to inward ulcers."

Hospitals are finding this to be true also. The General Hospital, Reading, impregnates all bandages for cuts, open wounds, etc., in honey, and a Norwich Hospital uses honey as a surgical dressing. There are doubtless other hospitals where the same thing is being done.

It has remarkable soothing power and eases the most excrutiating agony, and because of this is invaluable for burns. In February 1935 the *Alpenlandische Bienenzeitung* quoted a report from a man who said, "In the winter of 1933 I heated a boiler of about 35 gallons of water. When I opened the cover, it flew with great force against the ceiling. The vapour and hot water poured forth over my unprotected head, over my hands and feet. Some minutes afterwards I had violent pains and I believe I would have gone mad if my wife and my daughter had not helped me immediately. They took large pieces of linen, daubed them thickly with honey and put them on my head, neck, hands and feet. Instantly the pain ceased. I slept well all night and did not lose a single hair on my head. When the physician came to see me he shook his head and said, 'How can such a thing be possible?'"

Since reading that report I have used honey for minor burns, such as one gets when suddenly touching a hot stove, and have found honey to be remarkably efficacious. Usually burns leave scars, but these are reduced till almost invisible when honey is applied. I have never used honey for a cut, as even fairly deep cuts clot over almost at once and heal within a day or two.

This healing power is due partly—perhaps largely—to the hygroscopic power of honey; that is, the ability to attract and absorb moisture. This is so strong that honey can absorb moisture from stone jars, and even glass bottles which apparently are not in the least bit porous.

All germs must have moisture in order to live. When this is withdrawn, they die. That is one reason why intense heat kills germs. The spirochete, the syphyllis germ, for instance, dies within a matter of seconds if moisture is withdrawn. It is a wonder, therefore, that no one has used honey, both internally as well as externally, to cure that scourge.

The moisture content of honeys varies greatly. That which comes from fruit trees and the lime is usually gathered in a moist atmosphere and sealed with more water than other kinds; and this it is that causes the fermentation that gives it so distinctive a flavour and fragrance.

Don't imagine that honey that flows has more water in it than solid honey. Usually the reverse is so. The darker honeys contain more levulose (fruit sugar).

As mentioned earlier, one of the things that scientists have discovered about honey is that it creates hemoglobin in the blood. Hemoglobin is simply another name for the colouring matter in the blood, and is contained in the red-blood corpuscles. Your blood should contain about 5,000,000 red corpuscles and only about 10,000 white cells.

Hemoglobin is a compound of iron which attracts and combines with oxygen, and the red cells act as carriers of oxygen to the lungs. They are always dying and being renewed. They are destroyed in the liver and spleen and got rid of by means of the bile.

If they are not renewed at the right speed you get a greater number of white corpuscles in your blood and fall into what is known as an anaemic state.

Thus, honey is one of the great enemies of anaemia. It doesn't enrich the blood. Nothing does that. Blood is neither rich no poor. It either has the right constituents in it, or not. So-called " rich " blood is just as harmful as " poor " blood. Honey helps to maintain the right balance of hemoglobin and red-blood corpuscles.

Honey is a fine heart stimulant, and should always be preferred to stimulants like brandy, or spirits. Spirits have a bad reaction. They pep up the heart so that it works faster and drives more blood through the veins, giving the illusion that they provide energy.

But the energy derived from spirits is a false energy and the moment the effect of the alcohol wears off, the heart, left without the temporary fuel that caused it to work harder, is more tired than ever.

Honey, on the other hand, is both a food and a fuel. It has no unpleasant reaction. It does not make one feel utterly washed out when the effect has worn off, for it does not wear off. The sugar in honey goes straight into the blood and feeds the system.

In the past many doctors have used honey in heart cases. Dr. G. N. W. Thomas, of Edinburgh, wrote many years ago in *The Lancet* that honey had a marked effect in reviving and strengthening heart action and keeping alive patients whose forces were ebbing fast. In one case where a patient was in a critical stage of pneumonia, Dr. Thomas fed him constantly with honey till he had consumed two pounds in a few days. The honey brought about an early crisis, a rise in temperature and a strong pulse.

Sir Arbuthnot Lane, who advanced the cause of food reform strongly among the medical fraternity, had great faith in the use of honey as a heart and muscle stimulant, as an energiser, and a medicament to be taken in all cases of stomach trouble. He was the outstanding British specialist of his day in stomach ailments.

Honey is a tonic in the true sense of the word, which is derived from the Greek "tonikos," meaning "to tone, or intensify." It should be given when the heart is failing and the patient sinking rapidly.

Many heart sufferers benefit by sipping a glass of warm water in which a teaspoon of honey has been dissolved, just before going to bed. This not only assists the heart, but is a soporific.

Dr. H. A. Schuette of the University of Wisconsin stated that honey contains practically all the minerals comprising the human skeleton; especially iron, copper and manganese and dark honeys contain more of them than the lighter kinds. "Iron," he says, "is important because of its relation to the colouring matter of the blood. The hemoglobin which we build from our food has the power of carrying the all-important oxygen to the tissues of our bodies. Without iron content, hemoglobin would not have the power of holding oxygen.

" Copper has the power to unlock the therapeutic quality of iron in restoring the hemoglobin content of the blood of patients afflicted with anaemia."

Many physicians say that the lack of minerals in the soil in which some food crops are grown may be the cause of much present day illness. But bees have an unerring faculty for drawing their nectar from crops growing in the richest soils—preferably natural humus—and so transferring the salts in the soil to their honey.

Observers have noted that where there are adjacent crops, some grown in humus and others in artificial fertiliser, bees will always make for those grown in humus.

And if we follow the example of the bees, we can't go far wrong.

Honey builds bodily resistance to coughs, colds and other ailments. Not long ago Mr. J. M. Ellis wrote to a national bee-periodical: " I have not had a day's illness since pre-war days as a result of replacing bacon and eggs by oatmeal and honey.

" This confirms Simmons' statement of 50 years ago, that germs and microbes could be disregarded by those adopting a rational honey diet."

The Salvation Army experiment mentioned in *We Are What We Eat*, by A. B. Cunning, M.B., and F. R. Innes, M.B., is a noteworthy one and should be brought to the notice of all parents. Girls in a Salvation Army home who were fed on fish and chips, tinned meat, and margarine, were tired, listless and quarrelsome.

Their diet was changed to one of raw fruit, vegetables grown in humus, fresh, crisp salads, cheese, meat, eggs and plenty of honey, and in less than two years they were changed into happy, contented, hardworking people. It would be too much to claim that honey alone was responsible for this remarkable metamorphosis, but the advisory doctors thought so highly of honey as the ideal food to supply energy that they gave it a high place on the diet sheet.

I can vouch for the excellence of honey, which I ate first because I liked the taste of it, and not for any health-giving or medical virtue it might contain. It looms large in my diet, which unlike that of the Salvation Army girls, consists only of fruit, vegetables, legumes, nuts, cheese and milk.

HONEY FOR HEMORRHAGE

Some people are born with blood lacking in certain constituents, and when their limbs are cut, or even crushed, they bleed, either externally or internally. A test was made at the University Farm, St. Paul, U.S.A., in which honey was used for the prevention and cure of anaemia. The test was conducted on chickens and rats, as most such tests are, and not on human beings.

Purely by accident the researchers discovered that when chickens were fed on honey, it became difficult to draw from them the minutest quantities of blood needed for hemoglobin tests because the blood coagulated so rapidly.

This made the experimenters go one step further. They found that some chicks had a marked deficiency in vitamin K (koagulation factor), so fed them on buckwheat, alfalfa and honeys; and all chickens on this diet showed a marked anti-hemorrhage reaction.

Similar tests have not to date been carried out on human beings, but if they are and the results are as remarkable, it will be a great advance on any blood-clotting medium found so far. At present snake venom is about the only anti-hemorrhage medium known. It can be used only under certain conditions and then only by experts. If honey is found to be only half as effective, scientists will have made a big forward stride.

HONEY KILLS BACTERIA

A good deal has already been said about honey as an antiseptic. Dr. W. G. Sackett, Bacteriologist at Colorado Agricultural College, carried out a number of tests to determine whether honey would destroy various disease organisms, and found that all micro-organisms introduced into honey died within periods of a few hours, and at the outside, days.

Dr. A. P. Sturtevant, of the United States Bureau of Entomology at Washington, went even further. He based his experiments on the theory of the distinctive power of honey to absorb moisture from anything it comes into contact with, and decided that if honey took away the moisture from germ-laden areas, the germs would die.

He proved that the germ that causes dysentry dies within 48 hours; that which causes chronic bronco-pneumonia after the fourth day; and the germ that brings about typhoid and which may also cause peritonitis, after the fifth day.

These experiments, conducted by scientific men, substantiate the belief held for centuries, that honey is an enemy of disease. It is used all over the Far East as a cure for boils. Mixed with soap or other ingredients and applied to a boil, it brings it rapidly to a head and draws out all pus and moisture.

IDEAL FOOD FOR CHILDREN

Honey has been used as a food for children since time immemorial, all over the world. In the days before doctors condemned the pacifier as unscientific, mothers dipped pacifiers in honey and placed them in babies' mouths. But it was Dr. Paul Luttinger, of Bronx Hospital, New York, who made one of the first organised experiments with honey. He came to the conclusion that sugar was an irritant, and therefore harmful. It was a prime cause of worms. So he discarded altogether its use for infants. His first experiment was on 419 children to whom he fed from one to two tablespoons of honey in 8 oz. of feeding mixture. He also substituted honey for cod liver oil and orange juice.

In cases of diarrhœa he mixed one teaspoonful of honey into 8 oz. of barley water. He fed it liberally to infants in cases of marasmus (wasting disease), rickets, scurvey and malnutrition because honey is rich in proteins, mineral salts and vitamins—all lacking in refined sugar.

Most people do not realise it, but honey is richer in mineral salts than either human or cow's milk, and in parts of South-East Europe where the peasantry is singularly sturdy, with excellent teeth, it is given as an antidote to tuberculosis.

Dr. Luttinger found honey to be a good sedative, a cure for constipation (as well as diarrhœa) and to have good diuretic effects.

He was only placing on a scientific basis knowledge that has been known for centuries and has become incorporated into the medicinal lore of savage and semi-savage tribes. Among the Wa-Samia tribe of British East Africa, for instance, the only food that passes a mother's lips for several days after childbirth is honey and hot water. And, after circumcision, boys are given only honey and water for a week.

Only coffee has an equal diuretic effect—that is, it promotes the secretion of urine—but in coffee there are some 18 different drugs, as well as the well known caffeine and tannin. Honey can be given freely to many who dare not swallow coffee.

The French, incidentally, consider the sting of the bee to be a powerful aphrodisiac; but the Hindus go even further and say that in this respect honey is superior. When novices are preparing for priesthood they must abstain from meat, women, perfumes and *honey.*

In his excellent book Dr. B. F. Beck writes, "A bee-keeping friend of mine suffered from tuberculosis of the kidneys and was given up by two doctors 15 years ago. He got to eating honey and plenty of it and is today as peppy as a youngster."

Dr. Beck goes on to say that many of his patients were opera singers and to them he recommended a mixture of three parts of honey to one of compound tincture of benzoin; another mixture equally effective is 2 oz. honey, 1 oz. lemon juice; 1 oz. pure glycerine.

Honey and warm water is used all the world over as a drink for soothing sore throats.

HONEY FOR WARMTH

When one grows older, that is, into the eighties, it becomes more difficult than ever to keep the body warm. Whereas honey alone will not do this, it helps considerably. Dr. W. E. Crockett, of Boston, an athlete and strong man, was a believer in simple food and the constant use of honey. He was almost a vegetarian, though not quite, and after reaching the age of 80 astonished his friends by swimming across Boston Bay, taking a dip in the sea in mid-winter, walking 25 miles in just over six hours, putting a 30 lb. dumbell above his head 385 times (a most unusual form of exercise for an old man) and standing with his arms outstretched horizontally for half an hour!

Another grand old man who regularly ate honey was Dr. Fisher, father of a Canadian Prime Minister. At 90 he suffered from persistent cold feet, so used to slip into a blanket suit, put on snow shoes and take a two-mile run through the snow at nine each night. Also a most unusual form of exercise for a great-grand father.

After his nightly exercise, Dr. Fisher always slept like a log, for his feet were warm. Just before he went to bed, he drank a glass of warm milk and honey.

No one would honestly claim that honey alone would help you to do these things. You must have a strong constitution, sound heredity, live a sensible, active life, and eat and drink the right foods. And honey is one of those foods—one of the most important of them, at that.

Sir Henry Thompson, a famous physician, said that those who die early are usually men given to heavy meals in which there is plenty of meat. Not because meat itself is bad for a healthy man; this despite the fact that the writer is a vegetarian, though not, one hopes, a food faddist.

Meat has a tendency to produce acidity of the blood. If young and strong and living a hard, vigorous life, you sweat out most of the poisons from your body; but as you grow older this is impossible and the poisons take control. So, as you grow older, heed the advice of Dr. Josiah Oldfield and substitute honey for meat.

If people ate more honey, there would be less gastric and intestinal ulcers. Dr. Schacht, of Wiesbaden, claims to have cured many hopeless cases of internal ulcers by the use of honey—without operations.

Why have ulcers when you can have honey instead?

Have you noticed that women usually live longer than men? Take note of the centenarians when their ages are reported in the newspapers, and you will find that women greatly outnumber men. Because women usually eat less meat than men, and even decrease the quantity as they age.

If you are a great meat eater, reduce the quantity as the years pass by, and increase the amount of honey.

There are people who think that meat makes blood; so they must eat meat to live. One fat old woman I know who suffers from blood pressure would wail like an air-raid siren if you took away a chop from her over-filled plate. She is so full of blood that she feels giddy when she bends; yet the idea of giving half a pint to the local hospital fills her with horror. She feels she is going to die.

Yet, Edmund Eckardt, a champion blood donor, has earned his living during the last 25 years by supplying blood to the hospitals, and in three years saved 50 lives. He always takes a good deal of honey for breakfast; and every day eats 30 oranges and honey.

Another famous blood donor—also a professional—is 53 year old Willhelm Klein, of Frankfurt, who set up a world's record by giving 879 pints of blood in 1,300 transfusions in the last 15 years. He adheres to a sound health-giving diet with a high calorific value in which honey figures strongly as well as a special preparation of his made from soya beans called blutello.

HONEY FOR RHEUMATISM

That bee-stings cure rheumatism has been regarded as an old-wives' tale for centuries; but Dr. Heermann, of Kassel, Germany, wrote in a medical journal in 1936 that honey contains some of the neutralising acids found in the poison injected when a bee stings, and that it is every bit as good for rheumatism, atrophy of the muscles, nervous conditions, tuberculotic glands and other complaints, and is much pleasanter to take. Within recent years doctors have found that bees' wax also has curative properties and hot wax is placed on the hands and feet of arthritic sufferers, with excellent results. This is another instance of bringing ancient practice up to date, for in Hungary it has long been the custom to make a honey poultice and wrap it round a gouty big toe, after which the pain vanished in an hour or two.

Honey has tremendous rejuvenating powers, too, and for some 2,000 years has been accepted as part of the Kayakalpa system of medicine mentioned in Hindu books. According to *The Statesman*, Calcutta, one of the leading papers in India, Pandit Madan Mohan Malaviya took a course of Kayakalpa (which means literally a change of body and a new lease of life) at the hands of Tapsi Bisundas Udasi when he was 76. Pandit Malaviya had immense faith in the Hindu Shastras, which constantly refer to this form of treatment, but his mentor was not quite so sanguine.

On 16th January he entered a darkened chamber on the banks of the Ganges. Here he lived on a diet of milk, butter, honey and *aonla*, resting all day and meditating both morning and evening.

He emerged on the 28th of February a changed man. The wrinkles that made his face resemble a spider's web had gone. His gums had grown firm and there were signs of new teeth sprouting. His friends could not recognise this sprightly gentleman as the same bowed, ageing and emaciated figure who had delivered the convocation address at the University of Lucknow in December of the previous year.

Hapsi Bisundasji Udasi, who is a follower of the Guru (teacher) Nanak, has himself undergone the Kayakalpa treatment three times: the first when he was 75 at Paras-uramkund in Assam at the time of the First Burmese War in 1825; again about 50 years ago at Kotban in the Muttra district; and the last 13 years ago at Parson in the Etah district.

All this will seem a trifle far fetched to those who know nothing of the Yogis and their methods of rejuvenating the body.

It is as well to add that honey alone did not do the trick, but it formed an important part of the diet. And.

observers say that when Pandit Malaviya emerged from his treatment he looked no more than a young 60.

Once again modern medicine seems to be keeping up with ancient practice, for in 1950 Dr. Thomas S. Gardner, of East Orange, New Jersey, who has been trying to extend the period of life, said, " I'm working in five fields of research.

" For instance, vitamin B6 will play a large part in my experiments. It has been found in certain parts of meat products. (B6 is pyridoxine, also found in Brewer's yeast.)

" Helping foods of Nature—fruit and vegetables and that richest of sources, the royal jelly from honeybee hives.

" I am working with blood plasma, because in the blood is an agency which can help to renew youth as well as cells which grow old.

" When I am ready to start on people the treatment would best be begun when they are about 30 years old."

Shaw (G. B. S.) was another who lived to a ripe old age on a vegetarian diet in which honey played an important part.

Mr. Michael Bulman, Surgeon at the Norwich Hospital, read that the Egyptians dressed wounds with honey, and after trying it out has used it in preference to modern germ killers for healing large wounds and keeping them free from infection. He wrote in the Middlesex Hospital Journal: "This very simple substance is non-irritating, non-toxic, bactericidal, nutritive, cheap, easily applied, and, above all, effective."

HONEY FOR THE FAT

According to Dr. Beck, honey, when taken alone and not mixed with other foods, was considered an excellent reducer by the ancients, and though at first this appears illogical, it is in keeping with modern ideas and knowledge. " Fats and sugars are both carbon-containing and energy-providing foods which burn up by contact with oxygen and create energy."

Sugars burn more rapidly and produce energy more quickly than fats. When sugars—especially honey—are taken into the blood they cause rapid combustion and consume fats; and in any body that is slow to burn fat, as in the case of fat people, the process is assisted when honey or plain sugar is eaten. The same thing, of course, applies to sweets; and you will find that most habitual eaters of sweets are thin.

Diabetes is another disease in which honey may sometimes —not always—be eaten in place of sugar, which has a disastrous effect. Many doctors have tried it; some with success. The fact that it has been successful in many instances is good reason for the medical profession to institute research into the problem.

Honey is also the only sweetener that can be taken when one is suffering with worms; and is in fact used as a cure for worms in parts of India.

Dr. S. Z. Vazifdar, of Bombay, told H. J. Wadey, Editor of " Bee Craft," that he had great faith in honey as a remedy for cuts, sores and burns—particularly the last—in Indian villages where antiseptics are as likely to be taken internally as externally.

Some years ago the distinguished Australian aboriginal artist, Albert Namatjira, whose pictures are hung in a score of galleries and three of which are owned by Her Majesty The Queen, found that living on white man's food, such refined flour and white sugar, made him put on weight and affected his heart. When he reached 18 stone the quality of his painting fell off.

Doctors could do little for him, so he reverted to an aboriginal diet, cut out white flour and sugar, and ate several pounds of honey ants daily. Soon he was several stones lighter and his heart fit and strong once more. " I always feel well when I keep to my diet of honey ants," Namatjira explained to his doctor.

In Central Australia aboriginal women dig down into the ant tunnels with yam sticks, capture the honey ants and store them. The honey ant, unlike the bee, does not make a comb but uses the bodies of selected members of the tribe to serve as reservoirs. These ants are made to swallow honey till their stomachs distend to the size of a grape. When the aboriginals capture them, they just pull off the ants' heads and pop the rest into their mouths—like grapes—for they contain 100 per cent. honey! When on the move, searching for food, they eat little else, for honey gives them energy.

For 25 years the author suffered from hay fever every June. Attacks started in the first week of June and ended in the first week of July. Then he read Dr. D. C. Jarvis' "Folk Medicine" in which honeycomb is recommended for hay fever and chewed one teaspoon of honeycomb when in the throes of a severe attack. The hay fever vanished within seconds, and each time it recurred, the same simple remedy banished it. It is not enough to take honey; honeycomb is the remedy. Neither the author nor Dr. Jarvis can explain why, but it works!

CHAPTER SEVEN

HONEY vs. SUGAR

IN THE nineteen thirties a French doctor, Le Goff, made the astonishing statement that some 80,000 children died each year from the direct effects of artificial sugar; that is, the pure white crystallised sugar that adorns most breakfast tables. When he attended sick children he firmly refused to allow their parents to give them one grain of white sugar, which he termed poison, and insisted on the coarsest kind of sugar—almost black—containing the valuable minerals that are refined out of existence in white crystals; or honey and maple syrup.

Many other medical men; Schultz and Knott, of the University of Chicago, for instance; Dr. W. E. Deeds, and Dr. Charles G. Kerley say that many child-ailments are to be traced directly to the use of white sugar; in infants, red, irritated tongues, hunger spasms, heart burn and belching; in older children, colds, colitis, swollen tonsils, bronchitis, asthma, vomiting, rheumatism, chorea, eczema, hives.

The children of Britain were healthier than ever in their history during the Second World War, when sweet rationing was most stringent, for most sweets, except those made from treacle and honey, are composed largely of white sugar.

The Board of the National Research Council, U.S.A., stated early in the nineteen forties that white sugar was entirely lacking in protective vitamins and minerals and was a factor in unbalancing the diet of the nation. They felt they could not place it on the list of recommended foods.

But the powerful sugar interests in the States—The National Beet Sugar Growers' Association and the Great Western Sugar Co. of Colorado, had this decision reversed. Ody H. Lambourne, president of Lambourne & Co., largest sugar brokers in the world, while addressing the National Confectioners' Association on 2nd June 1943, denounced the scientists' findings as "government propaganda," and said, "Is it not reasonable to ask whether some sinister 'plan' exists whereby our people have to be withheld from them, by decree or otherwise, the foods which they desire?"

The unfortunate people both of America and Britain have little choice. It is possible, but difficult, to buy really good brown sugar in Britain; sugar that has not been devitalised by having all the valuable minerals taken out.

The British are in the position of Mr. Ford's customers in the nineteen twenties, who could have Ford cars of any colour they liked, provided they were black.

White sugar does not bear comparisan with honey.

Devitalised bread is at least "enriched" by the addition of synthetic thiamin and other vitamins and minerals, but no one has ever thought of "enriching" sugar.

Keep to honey, which is in every respect a superior food to white, or for that matter, brown sugar.

See that your bottle of honey is labelled "Pure Honey," and that it is not adulterated with sugar of any sort. The only sugar worth eating is "foot sugar," which is so dark and has such a strong tang that it takes all the taste out of tea.

CHAPTER EIGHT

SOME POINTS OF INTEREST

AS STATED EARLIER in this little book, there is no such thing to the connoisseur as " just honey." One might just as well tell a member of the Food Society that he is having " wine " with his dinner.

The expert apiarist will stick his nose into a jar of honey, take appreciative sniffs, look at the colour and texture, see whether it is proved or not and then dilate upon its virtues.

Ripe honey, for instance, is that which has been sealed over by the bees. The amount of levulose sugar exceeds the dextrose; the darker the honey, the more levulose it contains.

If ripe honey has less than 30 per cent. of water in it, it will keep indefinitely if sealed in airtight jars.

When unripe honey is extracted there is a tendency because of certain yeasts in it, to forment at temperatures between 50 and 140 degrees Fah.

The texture of honey varies greatly from a dark liquid that pours with the ease of castor oil, to a completely solid extract that has to be taken out with a spoon, or may be cut like fudge.

If kept, even liquid honey will granulate and the dextrose sugar separates into crystals. Light and temperature have an effect on granulation, and if a large crystal is wanted, honey should be allowed to granulate slowly, in good light and an even temperature.

The honeys of Ancient Greece and Palestine have been famous since time immemorial, but of modern European honeys, that made on the slopes of hills near Narbonne is among the most famous. It is white, granular and highly aromatic, being gathered from labiate plants, such as the rosemary. Some say that the honey from Corbieres, some 6-9 miles south is even better.

English honey made from the white clover, which is amber or even orange, is a food for the truly finicky. So is dark-golden heather honey.

Maltese honey is noted for its distinctive, delicate flavour, for it is gathered from orange blossoms.

The stingless bees of Brazil, too, produce excellent honey which varies widely, according to the flower from which it is brewed. It is highly rated, rather thinner than usual, but highly aromatic and considered to have a considerable medicinal value.

So for that matter has the eucalyptus honey of Abyssinia and Australia, from which country the trees were originally taken. It has a smell and taste of its own. It is from this brew that the Trappist monks of France make their celebrated " Liqueur de Tre Fontane."

So don't think of honey merely as a sweetener. Savour the different kinds and get to know them, and you can vary the honey you eat, just as you do jam. Every country, even the Arctic Circle, where bees work twice as long in the summer months, produces its own distinctive brew from flower nectar.

The Indian Government is training hundreds of farmers in bee-keeping and there are already 15 area offices, 150 sub-stations and 25 model apiaries in the country, and the number of official hives has increased from 17,500 to 50,000, and the number of bee-keepers from 4,000 to 10,000. Children in schools are to be taught bee-keeping as an additional subject and then to be sent to coffee, orange and banana plantations. The importance of the bee as a fertilising agent has been realised and the sum of 500,000 rupees has been allocated for training bee-farmers. (£1 equals approximately 13 rupees). (From *India News*).

CHAPTER NINE

HONEY RECIPES

ONE OF THE most famous makers of face creams and lotions uses a certain amount of honey in all skin preparations, because honey is an antiseptic, draws blemishes to the surface and dissipates them, and tightens the skin and causing wrinkles to vanish.

A good face cream can be made by mixing one dessert spoonful of honey into the well beaten white of an egg and adding a few drops of almond oil. Mix till a fine, smooth cream is formed.

SKIN LOTION

Mix—this can be done in a bottle by violent shaking—1 dessert spoonful of honey, 3 oz. glycerine, 1½ oz. lemon juice, ½ oz. red lotion, 1 oz. alcohol, 2 oz. rose water.

GARGLE

To 125 gm. of honey and 25 gm. of alum add one quart of water. Mix well. This is also excellent for sore throat and ulceration of the gums.

FOR COUGHS

Mix four tablespoons of honey, 1 teaspoonful of sulphur and 5 drops of pure turpentine. Take half a teaspoonful every two or three hours.

HONEY LEMON CHEESE

Mix half a pint of strained honey, the beaten yolks of three eggs, the whites of two, the juice of two lemons, the grated rind of one, and 1½ oz. butter.

Stir the mixture over a slow fire till it thickens. Pour into warm jars and seal. Honey lemon cheese will keep a year or more in a cool, dry place.

HONEY MARZIPAN

Ingredients: 1½ oz. margarine, 2 oz. sugar, 1 tablespoon of honey, 1 oz. cocoa, 2 tablespoons water, 4 oz. soya flour, and vanilla essence made from the vanilla pod.

Place margarine, sugar, honey and water into a saucepan and bring to the boil. Remove from the stove and add soya flour and cocoa, together with sufficient essence to give it a good flavour. Knead well until free from cracks. Then roll out until a quarter of an inch thick, brush lightly with jam and press on to your cake.

ROSE PETAL JAM

Ingredients: 2 breakfast cups of rose petals, 2½ cups sugar, 2 breakfast cups of warm water, 2 tablespoons of liquid honey, 1 teaspoonful of lemon juice.

Cut the petals into quarter-inch strips, discarding the rough base. Cover two breakfast cups of petals with 2 cups of warm water and cook for 10 minutes until they are tender. Strain petals and make a syrup with 1½ cups of the rose-petal liquid and 2½ cups sugar, and 2 tablespoons of liquid honey. Boil gently for 5 minutes, add the drained rose petals and cook on a low flame on an asbestos mat for 40 minutes. Add one teaspoon of lemon juice and simmer for 20 minutes more. Pour into hot jars and seal at once. A little carmine may be added to give colour—or cochineal.

Remember the old couplet:

> Of all the meals you can buy for money,
> Give me wholemeal bread and honey.

There is a good deal of truth in that.

In winter there is no finer drink when you come in from the cold than a teaspoonful of honey, the juice of half a lemon mixed into a glass of hot water. Add a few cardamom seeds to give a piquant flavour if you wish, a stick of cinnamon, or a few cloves.

An excellent salad dressing can be made with honey, lemon or lime juice and a spoon of olive oil.

When you bake apples, don't sprinkle them with white sugar. Core them and fill with either sultanas, raisins or dates—or a mixture of all—and when you take them out of the oven piping hot, pour a spoon of honey over them.

For breakfast an excellent dish known as muesli can be made with wholemeal flour (compost grown) or oatmeal soaked overnight in boiling water. Cover next morning with chopped apple or other fruit (and/or nuts), lemon juice and a dessert spoonful of honey. Add, if you like it, milk or cream.

HONEY AND NUT COLD CAKE

1 lb. stale wholewheat breadcrumbs
3 oz. margarine or butter or mixture of butter and margarine
Juice of one lemon
½ lb. clear honey
¼ lb. soft brown sugar (preferably foot sugar)
2 oz. dessicated coconut or chopped nuts

Melt the sugar, fat and honey together in a thick saucepan and boil for 2 minutes. Stir in breadcrumbs, lemon juice and nuts, beat well and leave to cool. Turn on to a plate or dish and shape into a round of square cake about 2-2½ inches deep. Put in the coldest place in the larder or in the refrigerator, till thoroughly cold. This cake is somewhat costly, but goes a long way and is rich enough to require neither icing or cream.

HONEY CAKES

1 cup wheat or oat flakes
1 tablespoon honey
1 tablespoon milk

Mix all well together, form into small cakes and serve with stewed fruit. Do not bake.

HONEY AND NUT CAKES

Mix 2 tablespoons of ground almonds with one tablespoon of orange or lemon juice and leave to soak for half an hour. Add one heaped tablespoon of wholewheat bread crumbs and one of honey. Mix, form into little cakes and serve.

HONEY AND ALMOND SANDWICH FILLING

Chop some almonds, add honey to the desired consistency, and spread between lightly buttered, thinly sliced wholemeal bread.

DATE BALLS

Chop ½ lb. stoned dates. Add 3 oz. chopped walnuts, or other nuts, 1 large tablespoon of honey, and 1 tablespoon of preserved ginger, finely chopped. Mix well, make into balls and roll in coconut or fine sugar.

AMBROSIA

Peel and slice finely some sweet oranges. Pour some honey over each slice and sprinkle with chopped cashew nuts, hazel nuts, almonds or pine kernels.

MUESLI

Mix together one tablespoon of chopped nuts or ground almonds, one tablespoon of rolled oats (or coarse meal) soaked previously in three tablespoons of water for 12 hours, the juice of half a lemon and a tablespoon of honey. Grate a large apple and add quickly to the mixture. Sprinkle a few chopped nuts on the mixture and a little cream or top of the milk, or yogurt.

PRUNE CREAM

Soak ½ lb. prunes till very soft. Mash them thoroughly, remove the stones and mix the pulp with 4 oz. honey. Place in a dish and sprinkle with dessicated cocoanut or milled nuts, and serve with junket, cream or yogurt.

HONEY PUDDING

4 oz. wholewheat bread
¼ pint milk
3 tablespoons honey
2 eggs

Cut bread into small cubes. Beat eggs with milk and one tablespoon honey, pour over cubes of bread and leave for 10 minutes. Grease a pudding basin, put the remaining honey in and pour in a mixture of bread, eggs and milk. If too thick, add more milk. Cover and steam for an hour and a half.

HONEY DESSERT

Mix to a thick consistency some finely chopped nuts and honey. Spoon into glasses and cover with top of milk, cream or yogurt. Garnish with glacé cherry or half a nut.

SHORTBREAD SHAPES

½ lb. self-raising flour
¼ lb. butter
½ lb. honey

If brown flour is used, Prewitt's Millstone is recommended

Work ingredients well together and mix in enough milk to make stiff dough. Roll out well. Cut into shapes and bake in hot oven till golden brown.

HONEY FUDGE

¾ lb sugar
¼ lb. honey
2 oz. margarine
½ cup milk
Pinch of cream of tartar
Dessertspoon of coffee essence

Slowly cook together sugar, honey, margarine and milk till sugar (Barbadoes or foot sugar) is thoroughly dissolved. Bring to boiling point, add cream of tartar and boil till the mixture thickens. Remove from heat, add coffee essence and beat well with fork till creamy. Pour into greased tins and leave to set.

HONEY POPCORN BALLS

1 lb. liquid honey
Popcorns or puffed wheat

Put the honey into a preserving pan, boil until it is very thick, stir in freshly popped corn, and when cool mould into balls.

HONEY BUTTERSCOTCH

Boil two teacups of honey until it hardens. Test by dropping a little into cold water. Stir the whole time to prevent honey burning or catching to the pan. Stir in half a cup of melted fat (butter preferably, of course, though margarine or nutter will do), add ¾ teaspoon of salt and flavour with vanilla (from the pod). Pour into cold, greased, shallow baking tin. When cool cut into squares and wrap in greaseproof paper to prevent the air making it sticky. This recipe may be varied by adding 2 cups of milk and boiling till the mixture is hard.

HONEY SALAD DRESSING

One part of lemon juice, 1 part liquid honey, 2 parts olive oil. Beat together and add the stiffly whipped white of an egg. Add—if you like—salt to taste.

HONEY MAYONNAISE DRESSING

1 egg
1 teaspoon salt
2 tablepoons honey
1 teaspoon vinegar
1 teaspoon mustard
6 teaspoons lemon juice
1½ teacups salad oil
¼ teaspoon pepper
A few grains cayenne

Break egg into bowl, add salt, honey, mustard, pepper, cayenne and one teaspoon vinegar. Beat thoroughly, add oil (one teaspoonful at a time), until half a cup is added and the dressing is thick. When one cup has been added, dilute with lemon juice, and at the same time add the remainder of the oil very gradually. Beat vigorously all the time while making.

HONEY TOFFEE

¾ lb. sugar	2 tablespoons honey
2 oz. butter or margarine	2 tablespoons milk
One dessertspoon of coffee essence	

Boil sugar, milk and fat until sugar is dissolved. Add honey. Boil rapidly until small quantity hardens on plate when tested. Remove from heat, add coffee essence, pour into well-greased tin and leave to set.

CHOCOLATE CRUNCH

1 oz. margarine	4 tablespoons honey
1 tablespoon cocoa	Pinch of salt
4 oz. rolled oats or barley flakes	

Warm margarine and honey and beat well together. Add cocoa and salt and beat again. Gradually work into the oast. Spread on a shallow greased tin (4 inches by 6) and bake in a moderate oven for 20 minutes. Mark into squares and cut when cold.

PEPPERMINT STICKS

1 lb. honey	5 tablespoons water
Pinch of cream of tartar or 1 teaspoon of vinegar	½ teaspoon of peppemint essence or a few drops of oil of peppermint

Slowly bring the honey and water to the boil and add cream of tartar or vinegar. Boil until a little of the mixture snaps when tested in cold water. Add peppermint and pour on to a greased plate. Leave until the edge takes the mark of the finger. Fold the sides of the mixture into the centre. Remove from the plate and pull mixture till it becomes lighter in colour. Cut into 12-18 pieces and pull into sticks. Leave to set on a flat, greased surface.

HONEY LEMON CHEESE

½ pint strained honey
Yolk of 3 eggs, whites of 2
Juice of 2 lemons and grated rind of 1
½ oz. butter

Stir the mixture over a slow heat till it thickens. Put into warm jars and seal. Will keep a year or more in a cool, dry place.

HONEY ICING

2 cups of powdered sugar
1 tablespoon melted butter
1 egg
1 tablespoon honey
1 tablespoon cream, in table-spoon vanilla

Beat the egg well, then beat other ingredients and make as smooth as possible. This will ice two cakes.

HONEY CREAM FILLING

2 oz. margarine
1 dessertspoon honey
A little warm water

Beat all well together. Add essence or cocoa to flavour.

STRAWBERRY OR RASPBERRY JAM

2 lb. fruit
Juice of 1 lemon
3 lb. sugar
¼ lb. honey

Mash the fruit and put into preserving pan on low gas. Add sugar, stir well until it begins to heat; add honey and blend thoroughly. Boil hard for four minutes, skimming meanwhile. Continue to stir for five minutes. Pour into *hot* bottles and cover while hot.

HONEY CAKE

6 oz. plain flour
3 eggs
2 oz. fat
2 oz. sugar
4 oz. honey
4 oz. dried fruit

Cream all ingredients, cook in a 5½ or 6 inch tin and bake at Regulo No. 4, reducing to 3—or 350 degrees, reducing to 300 in an hour.

HONEY QUEEN CAKES

2½ oz. margarine
1 oz. sugar
1 tablespoon honey
6 oz. self-raising flour
¼ pint milk and water
1½ oz. currants
Few drops vanilla essence
Pinch of salt

Cream margarine and sugar together till light; then add honey and essence and beat well again. Sieve flour; mix well and put into greased bun tins. Bake in a moderate oven for 20-25 minutes.

HONEY BISCUITS

2½ oz. margarine
1 oz. sugar
2 tablespoons honey
Pinch of salt
6 oz. self-raising flour
1 teaspoon cinnamon

Cream the margarine and sugar. Add honey, work in the flour, cinnamon and salt. Roll out until ¼ of an inch thick. Cut into rounds, place on a baking sheet and bake in moderately hot oven for 10 minutes. This quantity makes about 40-50 biscuits. If sweet biscuits are required, sandwich together with honey.

STEAMED HONEY PUDDING

½ cup honey
½ teaspoon ginger
6 oz. breadcrumbs
2 egg yolks
½ cup of milk
2 tablespoons butter
Rind of lemon
2 egg whites

Mix honey and breadcrumbs. Add milk, seasoning and yolks. Beat mixture thoroughly; then add butter and white of eggs well beaten. Steam for about 2 hours, filling bowl not more than three quarters full.

HONEY CHOCOLATE SPONGE

8 oz. self-raising flour
¼ teaspoon salt
3 oz. fine sugar
1 tablespoon honey
12-14 tablespoons hot water
3 oz. margarine
1 teaspoon bi-carbonate soda
Essence
(You may omit the bi-carbonate of soda if you wish)
2 tablespoons cocoa

Mix flour, salt, cocoa and sugar. Dissolve honey in water and add bi-carb. Melt fat and mix all ingredients till soft. Divide into 2 eight-inch tins and bake in moderate oven for 20 minutes. When cold, join with desired filling.

HONEY MALT LOAF

1 lb. flour
1 level tablespoon malt
½ level tablespoon honey
3 oz. fruit
½ level tablespoon syrup
½ pint milk and water
Pinch of salt
Use self-raising flour

Mix syrup, honey, malt, and milk and water over gentle heat; pour into fruit and flour mixture and mix well. Put into well greased tin and bake in moderate oven—Regulo 4—for about an hour. *Do not cut till next day.*

HONEY AND APPLE BATTER

Half fill a greased pie dish with sliced apple dipped in honey. Add tablespoon of water and trickle over more run honey. Pour over the whole a sufficiency of batter made at least an hour before, and well beaten before use. Bake for an hour in a moderate oven.

HONEY DROP BISCUITS

1 cup each of butter, honey, chopped nuts and sugar
1 egg
1½ cups plain flour
2 tablespoons baking powder

Cream butter and sugar. Add honey and beaten egg. Add nuts, flour and powder. Drop from teaspoon on a greased tin and bake in moderate oven for 15 minutes.

HONEY CREAM CHEESE

Mix one teaspoon honey with one of cream. Blend with a small cream cheese and serve on oat cakes, Ryvita or dry biscuits.

UNLEAVENED WHOLE WHEAT BISCUIT

4 cups wholewheat flakes
½ cup honey
½ cup nut butter dressing
3 eggs

Beat whites and yolks separately; then mix yolks, nut butter and honey. Now mix in white of eggs and finally wheat flakes. Drop a spoonful at a time on greased pans and bake for 15 minutes or until light brown.

UNLEAVENED FRUIT BREAD

2 cups water
1 cup cracked wheat
1 cup seedless raisins
½ cup stoned dates chopped fine
½ cup figs chopped fine
1 cup nut cream
½ cup honey

Soak the wheat overnight in water, add other ingredients in the morning and mix thoroughly. Steam in a double boiler for two hours, then shape into loaves of convenient size and bake in moderately hot oven.

WHOLE WHEAT RAISIN BREAD

1 lb. wholewheat flour
1 cup seedless raisins
1 tablespoon honey
1 tablespoon oil (salad oil)
1¼ cups water or nut milk
1 oz. yeast
1 teaspoon salt if desired

Dissolve yeast in ½ cup of warm water; add salt, honey, oil and raisins. Put in a pan and add flour and remainder of water. Knead well for 10 minutes and keep in a warm place for several hours. Knead again and put dough into well greased pans. Let it rise for an hour, then bake for 45 minutes. Brush top of loaves lightly with oil.

HONEY COCONUT MACAROONS

½ cup coconut oil or nut cream
½ lb. honey
1 cup hot water
½ lb. shredded coconut
¾ lb. wholemeal flour

Mix oil, honey and hot water thoroughly; then stir in shredded coconut and flour. Put in convenient shapes in a well greased pan and bake in moderate oven for half an hour. Should make about 3 dozen macaroons.

CRANBERRY SAUCE

2 cups cranberries
2 ripe bananas mashed
1 tablespoon nut crem
1 tablespoon honey

Wash and put cranberries through food chopper or mincer and mix with other ingredients. Stand for an hour before serving. This sauce is excellent for puddings.

MEAD

Recipe supplied by G. A. W. Collins.

Mead is the product of honey and water, fermented with yeast. Metheglin contains the addition of herbs and spices (Ground Gruit). Until recent years, much was left to chance; times have changed, however, and we can now rely upon selected ferments to do the work for us, with satisfactory results.

Cleanliness must be rigidly observed during all stages of production. Use only best quality honey, mix with cold boiled water; unless there are signs of fermenting, do not boil. If necessary, boil with the water for thirty minutes, making good evaporation loss with boiled water.

With the Extra Dry Yeast, use 3 to 3½ lb. Honey, standard blend; for Dry Mead 3½ to 4 lb, Medium 4½ to 5 lb., and Sweet 5½ to 6 lb.; this will be found the maximum amount needed for any brew. Excess is wasteful. All weights to one gallon of water, as above.

Preparing the Yeast: With a little tepid water and half a teaspoonful of sugar, mix into a smooth paste, and set aside for an hour, add to half a pint of the must, rest for twelve hours, then return to the bulk. Within twenty-four hours, at 65 to 70 degrees F., the first (rapid) stage of fermenting will commence; this easing, fit the fermentation lock (this is the foundation of success and is essential), storing at 65 degrees F. Working finished, disturb the cask; if no further action, bung lightly, tightening gradually daily for a week. Store at least nine months before racking; temperature 60 degrees F.

Racking: Into thoroughly well cleansed bottles, syphon or draw off through a tap the clear Mead until it appears cloudy, when together with the lees it may be filtered through a " Whatman's No. 1 rapid paper. Use new corks and fill to within half an inch of the base of same. Store at 55 to 58 degrees F., resting on sides, tie corks and seal with wax.

For Exhibition: From the bottles you have rested, pour one half or a third of the contents into standard round type " White " glass wine bottles, free from any marks or indentations. Store and treat as above.

Oak casks will be found the best medium for brewing and maturing Meads, Metheglin and Honey Wines.

No nutrients are required with the Yeasts; a special blend is available for Parsnip Wine.